ALPINE MEADOW

WEBS OF LIFE

ALPINE MEADOW

Paul Fleisher

BENCHMARK **B**OOKS

MARSHALL CAVENDISH
NEW YORK

The author would like to acknowledge the work of Paul Sieswerda of the New York Aquarium for his careful reading of the manuscript; Jean Krulis for her elegant design work; and Kate Nunn and Kathy Bonomi for their capable editing. He would also like to express deep appreciation for the loving, patient support that his wife, Debra Sims Fleisher, has provided for many years.

Benchmark Books
Marshall Cavendish Corporation
99 White Plains Road
Tarrytown, New York 10591-9001

Illustration by Jean Cassels

Library of Congress Cataloging-in-Publication Data
Fleisher, Paul.
Alpine meadow / Paul Fleisher.
   p.   cm.—(Webs of life)
Includes bibliographical references and index.
Summary: Examines the activities, plant and animal life, and climatic changes found in an alpine meadow in the Rocky Mountains.
ISBN 0-7614-0836-3
1. Mountain ecology—Juvenile literature. [1. Mountain ecology. 2. Ecology.] I. Title. II. Series: Fleisher, Paul. Webs of life.
QH541.5.M65F58 1999   577.5'38—dc21   97-38075   CIP   AC

Photo research by Ellen Barrett Dudley

Cover Photo: Animals Animals / Yva Momatiuk & John Eastcott

The photographs in this book are used by permission and through the courtesy of: *The National Audubon Society Collection, Photo Reserachers, Inc.:* Steve Coombs, 2; G. C. Kelly, 6, 12(top); Stephen Kraseman, 10; Renee Lynn, 12(bottom); David Marcias, 13; Carlton Ray, 14; Gregory G. Demijian, 16, 31(left); Joe DiStefano, 18; R. J. Erwin, 19; Stephen Dalton, 20; Scott/Jacana, 21; Anthony Mercieca, 23; Tom & Pat Leeson, 24, 30; Kent & Donna Dannen, 25; Adam Jones, 27; John Dommers, 28; Tom McHugh, 29; Gregory K. Scott, 31(right); Jeff Lepore, 32(left), 33. *Animals Animals / Earth Scenes:* Richard Day, 7; Robert C. Fields, 11; Jack Wilburn, 15; Robert A. Lubeck, 17; Jose Schell, 22; Alan G. Nelson, 26; Cathy & Gordon Illo, 32(right); David J. Boyle, 34.

Printed in Hong Kong

6 5 4 3 2 1

For Debra, with love

A yellow-bellied marmot suns itself on a lichen-covered rock. Butterflies flutter among thousands of blue, white, pink, and yellow flowers. A trickle of water gurgles beneath a slowly melting snowbank.

We are in an alpine meadow in the Rocky Mountains of Montana. It's a bright midsummer day, but snow still clings to the stony peaks.

7

Life in the meadow is harsh. At more than 8,000 feet (2,600 m), the plants and animals enjoy only about four months of warm sunshine. For the rest of the year, they must survive heavy snowfalls and extreme cold.

It is too cold and windy for trees to live in the meadow. The temperature can drop below freezing even in mid-summer. The winter wind blows hard enough to snap tree trunks and tear off branches.

We must treat the meadow plants carefully. The soil here is rocky and poor in nutrients. The harsh climate makes alpine plants grow slowly. If we damage the meadow by trampling or digging up plants, it takes a long time to recover.

DWARF WILLOW

To survive, alpine plants hug the ground. Lie down on the springy carpet of meadow plants. You'll discover that it's much warmer close to the ground. There is also less wind to dry out leaves and break branches.

Some alpine plants, like this willow, are low-growing versions of familiar trees or shrubs. In a warmer climate, the willow would grow much taller.

This plant is called old-man-of-the-mountain because of the many gray hairs on its leaves and stems. The hairs hold in warmth and moisture. The flowers of the old-man-of-the-mountain face south, to gather the most warmth from the sun.

OLD-MAN-OF-THE-MOUNTAIN

BIGHORN SHEEP WITH YOUNG

BULL ELK

The summertime meadow is a lush pasture for large plant-eating animals. Elk and deer come up the mountain to graze. When winter comes, the animals will move down from the high meadow to find shelter in the forest.

If we're lucky, we might spot a herd of bighorn sheep. In the fall the male sheep fight for mates, butting one another with their large, curled horns. Bighorn sheep were more common in earlier times. Much of their territory has been taken over by ranchers. Hunting and disease have also killed off many bighorns.

We might see mountain goats grazing in the meadow or picking their way along the stony cliffs. Goats are incredible climbers. Their shaggy white coats help them stay warm, even during the mountain winters.

MOUNTAIN GOAT AND KID

Small meadow plants can be many years old. Most alpine plants are perennials, meaning they stay alive year after year. It's difficult for plants in an alpine environment to sprout, grow, flower, and produce seeds in a single year. The growing season is too short.

It doesn't rain much in summer. The meadow plants' long roots reach deep into the ground to find water. They also store extra sugars and starches in their roots to give them a jump-start each spring.

SAXIFRAGE

ALPINE BISTORT

This food lets them grow leaves and flowers as soon as the weather turns warm, so they don't miss any of the summer.

Most plants produce seeds. But many alpine plants have found other ways to spread. Saxifrage sends out runners—thin stems with tiny new plants sprouting on the end. Alpine bistort grows little bulbs on its flower stalk. The bulblets will drop off and take root.

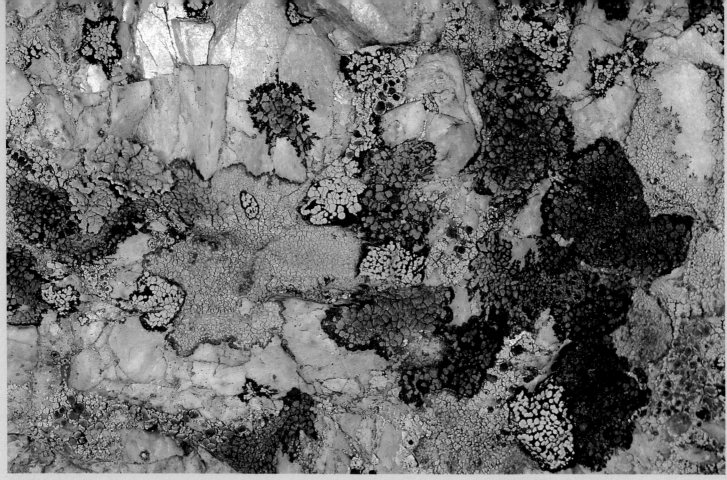

LICHEN

Patches of gray-green lichen grow on the rocks in the meadow. Lichen is actually two different things—fungus and algae—living together. The fungus clings to the rock and gathers moisture, but it cannot make its own food. Algae make food like other green plants. The algae share food with the fungus in return for a secure, moist place to live.

The snowbanks are pink in spots where watermelon lichen grows. It even smells like water-melon!

Reindeer lichen also grows among the meadow plants. Some grazing animals, like elk and deer, eat the lichen.

REINDEER LICHEN

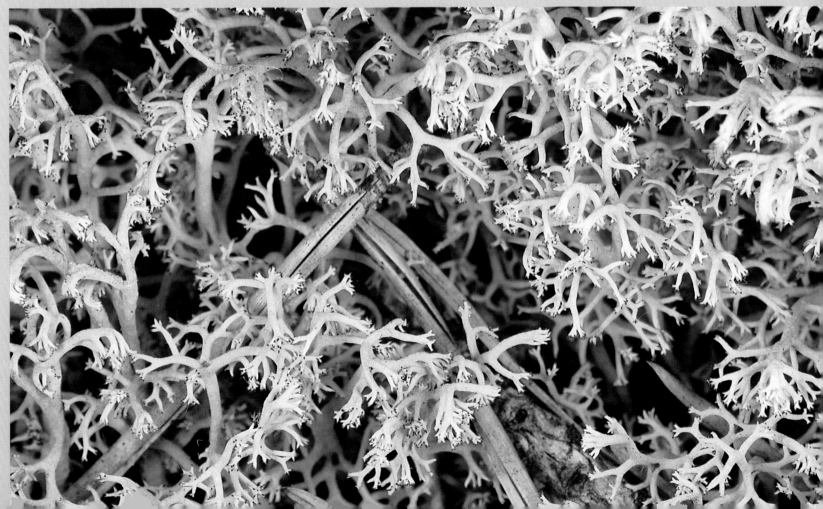

The plants in the meadow use energy from sunlight to make food. Some animals eat the plants directly. Others, called predators, eat animals that eat the plants.

The animals that live in the harsh alpine climate survive in a variety of ways. Some sleep through the winter or live underground. Others migrate, or move, to warmer areas. Insects lay eggs in winter that hatch only when the warm spring weather comes.

Kneel down and look closely at the meadow plants. They are buzzing and crawling with insects. As this bumblebee collects nectar, it carries pollen from flower to flower. The pollen fertilizes the flowers so that they will produce seeds.

BUMBLEBEE

HOVERFLY

Flies also pollinate alpine flowers. This hoverfly looks like a bee, but it has no stinger. The fly's beelike appearance fools birds, and protects it from being eaten. This form of imitation is called mimicry (MIM ick ree).

Like bees, butterflies such as this lustrous copper and Shasta blue flit among the flowers, sipping sweet nectar and spreading pollen. Butterflies lay their eggs on the meadow plants. The newly hatched caterpillars munch on the leaves.

Insects must be warm before they can move. Butterflies warm themselves by basking in the morning sun. Bumblebees heat their muscles by flapping their wings rapidly. Many alpine insects make a kind of chemical antifreeze in their blood that lets them keep moving when it gets cool.

SHASTA BLUE

HORNED LARK NEST

The birds and the mammals in the alpine meadow are all warm-blooded. That means they stay warm by burning food energy. Their inner heat lets them keep moving in cold weather.

Because there are no trees here, birds build their nests at ground level. The female horned lark scratches a shallow hole in the ground for her nest. She tries to hide her nest behind overhanging leaves. If a predator comes too close, the female leaves the nest to make it even harder to find.

ROSY FINCH

Rosy finches build nests in cracks or holes in rocky cliffs. For their nests, the finches make small, soft bowls of grass, moss, and feathers. They visit the meadow to feed. During the summer, birds find plenty of insects and seeds to eat.

23

GOLDEN EAGLE

A golden eagle's huge nest of sticks and branches perches on a rocky cliff. The eagle soars high above the meadow, searching for unwary marmots, ground squirrels, or other prey. The wings of this great bird are six feet (2 m) across!

When winter comes, most birds leave for a warmer climate. The white-tailed ptarmigan is the only bird that spends the whole year in the alpine meadow. The ptarmigan's summer feathers are brown on top and white below. But in winter, the feathers turn pure white. The ptarmigan scratches the snow in search of berries and plants. Its white feathers camouflage, or hide, it from predators.

WHITE-TAILED PTARMIGAN

Some mammals, like the marmot, survive all year long in the meadow by hibernating through the winter. Marmots spend about eight months of the year sleeping in their underground burrows. When they hibernate, their heartbeat and breathing slow down. Their body temperature drops to just a few degrees above freezing. They live on fat that they've stored in their bodies.

In the spring, the marmots awake from their long sleep. Within four months they must mate, raise their young, and eat enough to last them through the next winter.

MARMOT

GROUND SQUIRREL

Ground squirrels also hibernate. During the warm months, the squirrels eat lots of seeds and plants to store fat for the winter.

Other mammals stay active all year long. Meadow voles make runways through the low-growing plants. During the winter these mouselike animals travel along the snow-covered paths, gathering plants and seeds.

MEADOW VOLE

POCKET GOPHER

Pocket gophers carry food in little pouches in their cheeks. Each gopher digs its own system of tunnels beneath the meadow. Gophers move through their tunnels all winter long, eating the roots of meadow plants.

It's surprisingly snug under the blanket of snow. The air on the mountainside may drop to 30 degrees below zero (-35° C). But beneath the snow it's only a few degrees below freezing.

PIKA

The pika is a relative of the rabbit. Pikas live in stony areas near the meadow. Each pika spends the summer cutting and harvesting meadow grasses and other plants with its teeth. The pika carries the harvest to its rocky home. It makes one or two large piles of hay that it will eat during the winter.

The weasel is a skillful hunter. In the winter, it tracks small animals through the snow. Then it captures the meal with its sharp teeth and claws. In the summer, the weasel's coat is brown. But each winter it grows a coat of white fur that helps it stay hidden.

LONG-TAILED WEASEL IN SUMMER . . .
. . . AND WINTER

BOBCAT

Larger predators
hunt in the meadow
each summer. This
bobcat is stalking
nesting birds and
small mammals
among the low-
growing plants.
Coyotes find plenty
of animals to eat
here, too.

COYOTE

Grizzly bears include alpine meadows in their summer travels. Grizzlies are one of North America's largest and most dangerous predators. A large male grizzly can weigh more than 800 pounds (about 360 kg)!

The bears' menu includes lots of roots, leaves, nuts, and berries, but they also dig up and eat small mammals like marmots. And they are even fast and strong enough to chase down elk or sheep. In the winter, grizzlies spend most of their time sleeping in snug dens, lower in the mountains.

GRIZZLY BEAR

In early September, the short summer season comes to an end. The plants turn red, gold, and brown. Snow showers dust the meadow with white. Summer visitors such as elk and rosy finches move farther down the mountain. The ptarmigan and the weasel put on their white coats. Marmots and ground squirrels curl up in their underground burrows to begin a long sleep.

By October, the alpine meadow looks cold and lifeless. But beneath the snow, plants and animals wait patiently for the warmth of spring, when they will fill the meadow with life once again.

ALPINE MEADOW IN WINTER

Can you name the plants and animals in this alpine meadow?

Turn the page to check your answers.

Plants and Animals Found in This Alpine Meadow

1. yellow-bellied marmot
2. old-man-of-the-mountain
3. elk
4. mule deer
5. bighorn sheep
6. mountain goat
7. lichen

8. reindeer lichen
9. watermelon snow
10. bumblebee
11. hoverfly
12. copper butterfly
13. horned lark
14. rosy finch

15. golden eagle
16. ground squirrel
17. white-tailed ptarmigan
18. vole
19. pika
20. weasel
21. coyote

22. grizzly bear
23. fritillary butterfly
24. small apollo butterfly
25. forget-me-not
26. pasque

FIND OUT MORE

Behm, Barbara J. *Exploring Mountains*. Milwaukee, WI: Gareth Stevens, 1994.

Bradley, Catherine. *Life in the Mountains*. New York: Scholastic, 1993.

Cooper, Ann C. *Above the Treeline*. Niwot, CO: Roberts Rinehart Publishers, 1996.

George, Jean Craighead. *One Day in the Alpine Tundra*. New York: HarperTrophy, 1996.

Mariner, Tom. *Mountains*. Tarrytown, NY: Marshall Cavendish, 1990.

Simon, Seymour. *Mountains*. New York: Morrow Junior Books, 1994.

INDEX

ABOUT THE AUTHOR

In addition to writing children's books, Paul Fleisher teaches gifted middle school students in Richmond, Virginia. He is often outdoors, fishing on the Chesapeake Bay or gardening. Fleisher has made several visits to the alpine meadows of the western United States.

The author is also active in organizations that work for peace and social justice, including the Richmond Peace Education Center and the Virginia Forum.